CIRCUMSCRIPTION
HYPOTHESIS

ILYAN KEI LAVANWAY

Ilyan Kei Lavanway
CreateSpace
Cocoa, FL

ISBN-13: 9781505429329
ISBN-10: 1505429323

Published by Ilyan Kei Lavanway at CreateSpace
Cocoa, Florida, USA

Library of Congress Control Number: 2014922217

Dedicated to my brother, Aric.

ILYAN KEI LAVANWAY

CONTENTS

ILYAN KEI LAVANWAY

1 THE MEST CONTINUUM

I have been thinking about this for a number of years. I have alluded to the idea in at least two of my books: Earth Sink (2010) ISBN 9780976800439, and Intelligent Universe (2014) ISBN 9781494905910.

Albert Einstein's theory of relativity explains a macroscopic relationship between mass and energy and the speed of light. In this work, I am using the terms mass and matter interchangeably. All matter has mass.

There is no such thing as immaterial matter. If anyone has or ever will suggest such a notion, you can rest assured it is a false notion. If something has mass, it either has or can have some type of material manifestation at some level of refinement, even if it is so refined that it cannot be discerned by known means. And if something exists, it either has mass or a mathematical equivalent of mass.

Relativity breaks down at diminutive scales. It cannot account for phenomena of quantum physics. Quantum physics and relativity seem to be contradictory rather than mutually supportive. Why? What is missing? What bridges the chasm of inexplicability and satisfies the demands of both schools of thought? Is there one equation, as Stephen Hawking spent his life endeavoring to discover, that accounts for everything? Is there one whole within which all things can be explained?

The answer to the latter two questions is an unequivocal yes. The answer to the previous two questions has not yet been discovered in a quantifiable manner. The answer to the first question, why, is the subject of this work.

The simple answer to why relativity and quantum physics seem incompatible is that we have limited ourselves to a compartmentalized paradigm. I propose that matter and energy do not merely occupy and influence space. They are space. And they are time. There are four known manifestations of one whole, eternal, and infinitely abundant substance. Matter, energy, space, and time (MEST). They are all the same substance. Each one is simply a manifestation of how the others are organized and influenced, and shall we say, how they behave.

We might also consider that the whole of this MEST substance, or perhaps better stated, the whole MEST continuum, is divided into two overarching, mutually exclusive castes: that which acts with intelligence and that which is acted upon by intelligence. That which acts intelligently may endow some measure of autonomy to that which is acted upon, such that that which is acted upon may function in service to that which acts intelligently.

I am coining the term MEST continuum, because I do not know what else to call this whole of which I speak. I have only ever heard of each of its four known manifestations considered individually. Spacetime and spatiotemporal are as close as I have ever heard mention of more than one of the known individual manifestations being considered in an integrated or unified form. And no, I am not talking about the medieval concept of an aether.

Nor am I talking about the four classical elements: earth, water, air, and fire. However, some loose analogy could be drawn, if we were to think of earth as symbolic of matter or mass, and if we were to think of water as symbolic of space, air as symbolic of time, and fire as symbolic of energy.

2 THE CERTAINTY PRINCIPLE

I do not believe there is any such thing as a truly fundamental or elementary particle. Given any particle (or wave), there will always be a way to break it down into constituent parts. And if we were to break down the constituent parts into their respective constituent parts, iteration after iteration, we would eventually discover that we are looking at space and time organized on infinitely more finely resolved scales. Infinite resolution.

This seems to defy the Uncertainty Principle. The Uncertainty Principle states that the more precisely you account for one variable in a system, the less precisely you can account for other variables in the same system.

The Circumscription Hypothesis makes the Uncertainty Principle unnecessary. We could even go so far as to call a subset of the Circumscription Hypothesis the Certainty Principle. The Circumscription Hypothesis acknowledges the idea that it is indeed possible to account for any and all variables of a given system with perfect precision and accuracy, simultaneously.

The Circumscription Hypothesis further suggests that each and every system has influenced, does influence, or will at some point influence every other system to some degree, in a perfectly predictable and precisely observable manner, not merely in a statistically probable or improbable manner. It even goes so far as to acknowledge that somewhere, some intelligence has been, still is, and forever will be intimately aware of everything.

3 UNORGANIZED MATTER

While there is no such thing as immaterial matter, there is plenty of unorganized matter. At first glance, unorganized matter brings to mind images of dust or gases floating randomly in space. Could it also refer to space itself? I believe so. Stephen Hawking suggested that when our universe was newly created and was about the size of an atom, it was uneven and fuzzy.

It sounds to me like matter that was once unorganized was being organized. Out of that fuzziness and unevenness came marvelously well ordered cosmic structures such as galaxies, solar systems, stars, black holes, planets, and untold myriads of things we have yet to discover. From small and simple means come great things.

When we speak of things that seem to have come into being out of nothing, we have not acknowledged the possibility that the things that came into being were organized out of the fabric of space itself. It is well established that matter cannot be created or destroyed. It can only be transformed, rearranged, changed from one state to another, or one form to another, or moved from one place and time to another.

We know that matter can be converted into energy. Nuclear fusion and the mutual annihilation of matter and antimatter are two processes that come to mind when speaking of converting matter to energy.

Dare we speak of converting energy into matter? Five loaves of bread and two fish were multiplied to feed five thousand men. A barrel of meal and a cruse of oil were endowed with the capacity to auto-reconstitute for many days after the original supply should have been completely exhausted.

If living tissue can be organized from the elements of the earth, then why cannot the material that makes up planets and stars and everything else be organized from the MEST continuum? No intelligence needs to reinvent the wheel, so to speak. One of the characteristics of intelligence is the ability to rearrange or improve upon that which has already been organized at some level.

I am suggesting that while it is possible to organize space itself into matter, there is plenty of space that has already been organized into matter, and plenty of matter that has yet to be further organized into objects such as planets and stars.

Space is infinite. Space is eternal. It has no beginning or end. There is no space that is not acted upon or that does not pertain to some intelligent influence, and there is no intelligent influence that does not govern some volume of space.

Space did not come into being. What would have occupied the void filled by space if there were ever a time when space did not exist? Space.

Like matter, intelligence cannot be created or destroyed. Intelligence permeates and is integral to the entirety of space.

4 GRAVITY AND ENERGY

Energy is a function of mass, $E=mc^2$. Mass is a function of energy, $m=E/(c^2)$. Gravity is a function of mass, $F_g=G(m_1m_2)/r^2$. Here, c is the speed of light, F_g is the force of gravity, G is the gravitational constant, E is energy, and r is the distance between mass one, m_1 and mass two, m_2.

Gravity is an influence, a force which acts. Is gravity intelligent? I don't know. I believe, at the very least, that gravity is one result of mass, mass being intelligently organized spatiotemporal energy, being endowed with some degree of autonomy such that mass, through the influence of the gravity, or shall we say the spatiotemporal curvature, with which it is endowed, acts in service to some intelligence. Gravity is curved space. Curved space is an expression of mass and energy.

How are mass and energy, and thus gravity, related to time and space? If gravity is a function of mass, and mass is a function of energy, then gravity, or spatiotemporal curvature, is a function of energy, $\mathbf{F_g = G(E_1E_2)/(c^4r^2)}$.

Consider that energy and mass are related by light. In other words, light is energy. Energy is light. Light, then, is a function of gravity, and gravity is a function of light. Light is electromagnetic energy. That means gravity is somehow related to electromagnetism. Gravity, electricity, and magnetism are somehow interrelated on the MEST continuum.

Okay, I know what you're thinking. Einstein's equation speaks of the *speed* of light, not light itself. It assumes the speed of light is constant within a given medium. A given medium? So the medium, that is the material through which light passes, influences the speed of light? A medium is something, thus a medium has mass, or a mathematical equivalent of mass somewhere on the MEST continuum, even if that medium is space and its mass equivalent cannot yet be measured. Of course, light is a transverse wave and does not affect the medium of space. Or does it?

5 TIME DILATION

Does the fact that light behaves as both a particle (matter) and a wave (energy) suggest that the MEST continuum is one whole that can behave as matter, energy, space, and time? If we can speak of the dual nature of light, can we not also speak of the quadruple nature of all things? Does not the quadruple nature of all things call to mind literal and symbolic images of mutually perpendicular axes? But I digress. My point, here, is that light is energy.

Within a frame of reference where light (energy) is infinitely concentrated, there is no time. An agent within an environment of infinitely concentrated light would be able to perceive absolutely everything that has ever occurred, everything that is occurring, and everything that ever will occur. All things are present in such an environment.

In other words, where light is infinitely concentrated, time is not a factor. Within such a frame, mass, energy, and space are organized and function independently of time. The only thing imperceptible, or irrelevant, in such an environment is time. Time becomes imperceptible, irrelevant. But, wait. If time becomes imperceptible or irrelevant, does distance, or space, also become irrelevant and effectively imperceptible? We'll get to that.

As you distance yourself from an environment of infinitely concentrated light, time immediately becomes a factor of increasing influence. Time becomes increasingly perceptible and relevant. The influence and the perception of time increase with distance as you move away from an infinite concentration of light. So, as you distance yourself from such a frame, you become increasingly aware of, subject to, and limited or bound by distance, spacetime. You become spatiotemporally challenged.

It is already well established that space curves more and more tightly with proximity to a singularity or any concentration of mass. As space curves more tightly, time seems to slow down when viewed from afar. For an agent approaching a singularity and looking outward, it would seem that time at his departure point accelerates as he draws nearer to the singularity. In other words, as the agent looks outward, he sees his friends and family aging faster and faster.

When he looks at his own wrist watch, he sees it clicking along at the same pace to which he has been accustomed. Nothing changes for him. But when he looks at his friends and family far away, he sees their clocks running faster and faster the closer he gets to the singularity. This is called time dilation. When his loved ones look at him, they see him aging more slowly, and they see his watch run more slowly the closer he gets to the singularity.

6 MASS DILATION

A similar phenomenon, called mass dilation, occurs when a mass is accelerated to speeds approaching the speed of light, relative to an observer. At a relative speed equal to that of light, an observer would perceive the mass as an infinite mass shortened into a plane perpendicular to the velocity of the mass.

If we visualize the mass from a point of view along its velocity vector, we see that the apparent cross sectional area (size) of the object remains proportional to its distance from our position. If it is coming directly at us, it appears to get larger in size. If it is moving directly away from us, it appears to get smaller in size, just as we would intuit.

The Doppler effect is in play to an extreme. To see the mass coming at us at light speed, we would have to perceive infinitely short wavelengths of the spectrum, because the object would appear infinitely blue shifted (I guess that's why Super Donkey is blue). To see it moving away from us at light speed, we would have to perceive infinitely long wavelengths of the spectrum, because the object would appear infinitely red shifted.

If we visualize the object from some location not along its velocity vector, and if we could view the object as it passes directly abreast of us, it would appear infinitely shortened, like looking at a flat sheet of paper edgewise, but so much thinner it would seem imperceptible unless we could perceive things on an infinitely refined scale or at infinitely slowed time. But, to do that we would have to be moving right along with the object, or we would have to be viewing the object from a frame of reference in which energy is infinite and time is stopped and space is infinitely compressed.

7 COLLOCATION OF EVERYTHING

In such a frame, distance (space) would in effect become imperceptible or irrelevant. In other words, to an agent within such a frame, it would seem as though all things were collocated with him, drawn into his presence. This might be perceived as the equivalent of being everywhere at once, or being able to go anywhere instantaneously.

Where there is no perception of time, there is no perception of distance. That does not mean distance, or space, ceases to be. It simply means space, distance, no longer prevents an agent from traveling to any number of locations instantaneously.

To understand the ramifications of this, consider the fact that in our mortal frame of reference it is virtually impossible to perceive all things as they really are. There does exist a frame of reference where all things are perceived as they are, as they have been, and as they will be. In such a frame, access, manipulation, and organization of mass, energy, space, and time are literally at your fingertips, and any amount of work or movement upon such can be accomplished with a thought.

If such a frame of reference were to exist within your very being, that is to say, if you were so filled with light that you were to become the equivalent of a luminous singularity, then you could go anywhere instantaneously, and retain omniscience and omnipotence. You would no longer be dependent upon or bound to any location or material.

Okay, an agent so filled with light as to become, in effect, a luminous singularity, can see everything, past, present, and future. How does that give him access to, and influence over, things actually far distant from him? After all, while he perceives all things as if they were in his immediate presence, there are actually vast distances between the agent and all else.

An observer far, far away would perceive the vastness of open space and the great separations between things, and wonder how an agent in the frame of a luminous singularity can do anything other than sit there and watch passively as stuff happens. How does the agent connect with that which he so effectively perceives? Wormholes.

Such an agent must be capable of generating wormholes at will. He must also be able to avail himself of the myriads of existing wormholes that connect everything to everything else. There must be a network of wormholes, or some equivalent thereof, permeating the cosmos. Is it possible that at infinite concentrations, or at infinite energy levels, each photon of light becomes, in effect, a wormhole, or is composed of infinitely many infinitesimal wormholes?

But, wait. Wouldn't it require virtually infinite amounts of energy to generate wormholes between places (or times) infinitely far removed? Probably. But maybe not.

There may be a simple means of generating wormholes with a small initial exertion. Likely, there is at least one way to set off some type of chain reaction or self-sustaining resonance in the MEST continuum that opens and controls wormholes.

Either way, an agent in a frame such as a luminous singularity would have infinite energy at his disposal. His very thoughts would be infinitely energetic and powerful. What are thoughts, after all? Electromagnetic energy acting intelligently. And where are these thoughts taking place? In an environment of infinitely concentrated energy.

Things that make you go, "Hmm ...," or jump out of your seat and shout, "Gnarly!" But, let's not get ahead of ourselves.

8 MASS MIMICS MOTION

We often hear of mass dilation and its accompanying reduction in length when speaking of traveling at relativistic velocities. However, mass dilation and reduction in length must also occur with mere proximity to a singularity. In other words, as an agent approaches a singularity, his loved ones observing from afar will perceive him to be increasing in mass. No, he will not look bigger, but he will seem more and more dense.

If he is approaching the singularity in a direction normal (perpendicular) to the curvature of its event horizon, in other words he is not orbiting the singularity, and if he could be viewed tangentially to the curvature of space at his distance from the singularity, would he appear thinner and thinner the closer he gets to the singularity? Would his body appear to curve more and more tightly? If I understand correctly, he would indeed appear thinner and more tightly curved the closer he gets to the singularity.

When viewed normal to the curvature of the singularity's event horizon, he would only appear to change size in proportion to his distance from observers afar, and he would be red shifted to infinitely long wavelengths until he would vanish from perception. Unless his loved ones viewing from afar have the ability to discern infinitely long wavelengths of light.

Red shift will occur with mere proximity to the singularity, even if the agent is stationary. At any given location, the gravity of the singularity produces the same effect as some constant velocity away from an observer.

It seems this case mimics the case in which an agent is approaching the speed of light and traveling into open space, directly away from observers. As he approaches the speed of light, he is red shifted. But what about his apparent size in relation to his distance from observers?

At a constant departure velocity, he would seem to shrink at a constant rate, proportional to his velocity. If he were accelerating away from his loved ones, he would appear to shrink at an exponential rate, proportional to his acceleration.

As his distance from observers increases, he would appear proportionally smaller. Does that mean he would also appear proportionally smaller with increasing proximity to a singularity? To put it another way, does a constant velocity directly toward a singularity make an agent appear as though he were accelerating away from observers afar? It seems that such would be the case.

This leads me to believe that acceleration away from an observer into open space and constant velocity toward an infinitely concentrated mass or infinitely concentrated energy are indistinguishable conditions. An agent may be stationary near a singularity and appear as though he had traveled an almost infinite distance directly away from observers afar. An agent may move at a some constant velocity directly toward an infinite concentration of mass or energy and appear as if he were accelerating away from observers afar.

So, there appears to be a sensitivity factor that becomes increasingly influential the closer an object or agent gets to an infinite concentration of mass or energy. The closer one gets to a singularity, the more a very small change in distance toward the singularity mimics a very large change in velocity away from observers afar. The closer one gets to a singularity, the more a very small change in velocity toward the singularity mimics a very large change in acceleration (a surge) away from observers afar.

One way to determine if an agent near a singularity is stationary along a line of sight normal to the curvature of space at the agent's distance from a singularity is to notice whether or not his apparent size is changing. If he is red shifted but does not appear to change size, he is stationary, and his distance from the singularity may be determined by the degree of red shift. If his apparent size is changing, or if the degree of red shift is not constant, he is not stationary.

A similar technique is used for determining the velocity of luminous bodies relative to Earth. The degree of red shift or blue shift measured against the assumed original wavelengths of light emitted from or reflected by the object allows velocity calculation.

We have been talking about the red shift that results from proximity to a singularity, or any significant concentration of mass or energy. Is there an opposite set of conditions that would cause an object or agent to appear blue shifted with mere proximity to some reverse or negative concentration of the MEST continuum, or some sort of bubble in the MEST continuum?

Is a negative singularity possible? Does negative mass cause blue shift and convex spatiotemporal curvature, as opposed to the concave curvature we commonly associate with positive mass? Would these conditions be present with a white hole? Or the moment immediately preceding the birth of a universe?

9 ONE ETERNAL ROUND

Another interesting phenomenon occurs when approaching the speed of light, or when approaching a singularity. If you were able to do either, you could look straight ahead and see not only what is directly in front of you, but everything abreast of you. This would progress until you could look straight ahead and see the back of your head and everything behind you.

Why? because space is becoming more tightly curved. Or you are in North Dakota where everything is so flat you can jump up and see the back of your head, and there's a pretty girl behind every hill, but I digress.

Space becoming infinitely curved, and thus bringing everything into view simultaneously, is consistent with all things being present within a frame of infinitely concentrated mass, energy, or light. An agent within a frame where space is infinitely curved is not subject to, or bound by, time or distance or space, or the speed of light. He can be anywhere instantaneously. That means, in effect, he can be everywhere at once.

In other words, he, or his influence, can be in all things and through all things. His course, then, is one eternal round.

CIRCUMSCRIPTION HYPOTHESIS

10 ELECTRIC UNIVERSE

While pondering these things, the theory of the electric universe makes more sense as it claims that only a fraction of the energy emitted from stars is generated by fusion within the star itself. If fusion were the sole mechanism of a star's energy production, the star would quickly fuse all its lighter elements into iron, after which fusion would become endothermic and the star would collapse. Left entirely to itself, it would not last long enough for planets orbiting it to mature and fulfill the intent of their creation.

The Electric Universe Theory suggests that stars are focal points for energy generated somewhere else. I believe one explanation may be that their mass and the concentrated magnetic fields, in conjunction with heretofore undiscovered quantum phenomena associated with the fusion process, enable a channel, or dare I say wormhole, through which the bulk of radiated energy, or perhaps supplemental incoming subatomic matter to be fused, flows from a much more powerful and uninterruptable source. In this sense, stars are analogous to light bulbs.

A light bulb emits light generated by electrons (incoming subatomic material) rushing through a metal filament, or through a gas, heating or ionizing it according to specific laws of physics under which it was designed to operate. The light bulb is a focusing mechanism for a tiny fraction of the tremendous amount of energy, or electrical potential, generated at a distant power plant and transmitted through a channel, a set of power lines and substations, an infrastructure, to the location where it is intended to illuminate a certain volume of space for a certain purpose. Such purpose often involves providing convenience to some form of life, including human life.

Another way to look at this may be to consider the whole of the cosmos, the whole MEST continuum, as an infinite reservoir in eternal flux, as if it were alive. I believe there are infinitely many spatiotemporal dimensions. We are only aware of three, described by three mutually perpendicular axes which we use to describe three-dimensional volume. Some argue that time may be a fourth dimension. I do not think it is. I think time can exist within any and all dimensions, except perhaps dimensions less than three. More on that later. Time may even have infinitely many dimensions, but it is not necessarily a dimension itself.

Okay, so back to the Electric Universe and how stars get energy from sources other than their own fusion. Is it possible that a significant portion of the energy radiated from stars is a result of matter-antimatter mutual annihilations giving off gamma radiation that is attenuated within dense regions of the stars, or attenuated as a result of heretofore unknown spatiotemporal phenomena at work within the stars? If so, where does the antimatter come from? Does the star generate it and then use it in some self-sustaining, controlled perpetual energy production process?

Could matter and energy be tucked away in other spatiotemporal dimensions that can be tapped, like infinite wells, by processes we cannot yet understand? Why not?

11 ROW, LAYER, STACK

I believe one characteristic of any given dimension is that it can be described by an axis that is mutually perpendicular to the axis or axes describing every lesser dimension. This is independent of time.

An object in one-dimensional space, a line, can move in only two directions along the axis, or line, that describes its space. If it moves at subluminal speeds and has a finite mass, relative to an observer, it is subject to time.

However, relative to a fixed observer, is it even possible for anything to exist in less than three-dimensional space? That brings up a whole new question: Is the number of dimensions in which an object resides relative?

Relative to a fixed observer, is it possible for anything in less than three-dimensional space to have finite mass and thus be subject to time? It seems to me that Relativity says no. More on that later. I believe Planck's constant has something to do with it.

Anyway, when infinitely many one-dimensional spaces are arranged adjacent to each other, or when one of them is rotated through infinitely many angles around some perpendicular axis, they collectively describe and constitute a two-dimensional space, a plane. Objects within a plane can move in infinitely many directions within the plane. It is questionable whether such objects are subject to time within their plane, given that a plane has less than three dimensions. But let's not digress to that yet.

When infinitely many two-dimensional spaces are stacked upon each other along an axis perpendicular to the plane, or when one of them is rotated through infinitely many angles about one axis within the plane, that axis being perpendicular to the other axis within the plane, the resulting set of planes describes and constitutes a three-dimensional space, a volume. Objects within a volume can move in infinitely many directions within that volume, and they can be subject to time within that volume.

If the same reasoning holds when infinitely many three-dimensional spaces are nested, one inside another, or arranged in a row, like a row of boxes, or when one of them is rotated through infinitely many angles about an axis that is mutually perpendicular to the three axes that describe the volume, the resulting set of volumes describes and constitutes a four-dimensional space. Objects within a four-dimensional space can move in infinitely many directions within that space, and can be subject to time within that four-dimensional space.

An agent capable of operating in four-dimensional space can view all faces of a three-dimensional object at once. Just like we can see all edges of a two-dimensional object at once.

Objects in n-dimensional space can move in infinitely many directions within that n-dimensional space, and can be subject to time within that n-dimensional space.

It is impossible for us, well, for me anyway, to visualize an axis mutually perpendicular to the three axes that describe three-dimensional space. Perhaps we can get away with visualizing four-dimensional space as a row of three-dimensional spaces arranged adjacent to each other, like a row of boxes.

After all, one-dimensional space, let's call it one-space, is a row of points forming a line. Two-dimensional space, two-space, is a set of lines arranged adjacent to each other, forming a layer, a plane. Three-dimensional space, three-space, is a set of planes stacked on top of each other, forming a cube.

So, it makes sense that four-dimensional space, four-space, is a set of cubes arranged in a row, like one-space is a set of points, zero-dimensional spaces, arranged in a row. Looks like patterns are beginning to emerge here.

It appears that we can visualize n-dimensional space, n-space, as a set of $(n-1)$-dimensional spaces, $(n-1)$-spaces, arranged in a row, or a layer of rows, or a stack of layers.

Notice, here, a pattern within a pattern is manifesting itself in sets of threes.

If we start in a space composed of a row, and we add a dimension, we get a space composed of a layer of rows. If we add another dimension, we get a space composed of stacks of layers.

If we add yet another dimension, we get a space composed of a row of those stacks. Adding another dimension gives us a space composed of a layer of those rows of stacks. Add yet another dimension and we get a space composed of stacks of those layers of rows of stacks.

We advance from a row of spaces to a layer of rows to a stack of layers, and then another row, and then another layer and then another stack, on and on, iteration after iteration, always repeating in sets of threes.

This is analogous to musical notes arranged in octaves. There are infinitely many octaves. Each octave is a set of the same notes expressed at lower or higher frequencies.

For any positive integer value of n, we can visualize $(3n-2)$-space as a row of $(3n-3)$-spaces. And we can visualize $(3n-1)$-space as a layer of $(3n-2)$-spaces. And we can visualize $(3n-0)$-space as a stack of $(3n-1)$-spaces.

If this reasoning holds, we can visualize five-dimensional space, five-space, as a set of four-spaces arranged adjacent to each other, forming a layer of rows of boxes. Picture five-space as a layer of four-dimensional spaces covering a warehouse floor.

Six-dimensional space, six-space, is simply a set of five-dimensional spaces stacked on top of each other. Picture an infinitely large warehouse filled with a stack of five-spaces.

I suppose we could also look at this like a Rubik's Cube. The main cube is like a piece of six-dimensional space. It is composed of layers of rows of smaller cubes. Each small cube represents a three-dimensional space. Each row of small cubes represents a four-dimensional space. Each layer represents a five-dimensional space. The stack of layers represents a six-dimensional space.

Note how this arrangement is analogous to three-space being a stack of two-spaces, while two-space is a layer of one-spaces, and one-space is a row of zero-spaces, points.

The pattern continues with seven-space being an infinitely long row of six-dimensional spaces. Picture a row of warehouses. Each warehouse is filled with a stack of five-dimensional spaces.

Eight-space is an infinitely expansive layer of seven-dimensional spaces. Picture eight-dimensional space as a layer composed of rows of warehouses.

Nine-space is a stack of eight-spaces. Picture an even bigger warehouse stacked with layers of warehouses. Here, the bigger warehouse is filled with a stack of eight-dimensional spaces.

Let's beat this dead horse a bit more and run through a few more iterations. We are going to arrive at a couple of observations, or rather, a couple of questions.

Ten-dimensional space can be visualized as a row of nine-dimensional spaces. Imagine a row of the bigger warehouses.

Eleven-dimensional space is a layer of ten-dimensional spaces. Imagine a layer of rows of the bigger warehouses.

Twelve-dimensional space is a stack of eleven-dimensional spaces. Imagine a ludicrously colossal warehouse filled with a stack of eleven-dimensional spaces. Observe that this does not stop at ten or eleven spatial dimensions.

Time can flow in all spatial dimensions, except perhaps in dimensions less than three. So, why do some mathematicians and cosmologists propose that there are only ten spatial dimensions, plus time? I don't know. Surely, there are mathematical relationships we have yet to discover.

Observe that, starting with two-dimensional space, every third additional dimension is a layer of the next lower dimensions. Two-space is a layer of one-dimensional spaces. Five-space is a layer of four-dimensional spaces. Eight-space is a layer of seven-dimensional spaces. Eleven-dimensional space is a layer of ten-dimensional spaces. Fourteen-dimensional space is a layer of thirteen-dimensional spaces, and so forth.

Dimension	Arrangement	Membrane
0-space	Point	No
1-space	Row	No
2-space	**Layer**	**Yes**
3-space	Stack	No
4-space	Row	No
5-space	**Layer**	**Yes**
6-space	Stack	No
7-space	Row	No
8-space	**Layer**	**Yes**
9-space	Stack	No
10-space	Row	No
11-space	**Layer**	**Yes**
12-space	Stack	No
13-space	Row	No
14-space	**Layer**	**Yes**
15-space	Stack	No
16-space	Row	No
17-space	**Layer**	**Yes**
18-space	Stack	No
19-space	Row	No
20-space	**Layer**	**Yes**
21-space	Stack	No
22-space	Row	No
23-space	**Layer**	**Yes**
24-space	Stack	No
25-space	Row	No
26-space	**Layer**	**Yes**
27-space	Stack	No
28-space	Row	No
29-space	**Layer**	**Yes**
30-space	Stack	No
31-space	Row	No

Layers. Membranes. Are we touching Membrane Theory, M-Theory? Or am I just touching fabric because I'm so engaged in my thoughts that I don't want to get out of my seat to red shift some three-dimensional mass into the porcelain singularity. But I digress.

So far, we mortals can only navigate in three-dimensional space. And we can only perceive things as two-dimensional representations. We only see one face of things at any given moment and from any given vantage point. There are infinitely many movements and processes that we cannot visualize or comprehend, yet.

12 PROTON UP CLOSE AND PERSONAL

In a dream I had more than a decade ago, I saw a close-up view of a slightly hazy, grayish-colored sphere loosely resembling a ball of tightly interwoven elastic threads in constant relative flux. The surface, if you could call it that, for there was no distinct solidity, was enshrouded in a translucent, gray haze.

The haze seemed like some type of an atmosphere. There was no distinct transition from atmosphere to surface, but the density of the gray haze seemed to increase exponentially with depth.

The denser topography had a stringy, swirling appearance, as if the entire sphere were a violently dynamic, fluid body with powerful storms resembling elongated, elliptically shaped typhoons with starkly delineated shear zones between concentric currents.

In the gray haze, at some depths, there seemed to be violent tornados. These tornado-like phenomena were extremely tightly wound and spinning at extreme speeds at the deeper ends. Toward the upper ends, they fanned out smoothly, swirling more slowly.

The dynamics of the entire sphere were such that the sphere itself did not maintain a truly spherical shape, but was more akin to a water balloon sloshing about. It never stayed in one shape, and never duplicated the same shape.

Somehow, I intuitively understood that this dynamic sphere represented a single proton. It was immensely magnified and displayed in slow motion to reveal intricate details and dynamics.

The proton was being bombarded with light in the form of individually discernable photons. I watched as individual photons of light impacted the proton and were absorbed into the proton.

At first, nothing seemed to happen. But, a few moments into the continuing barrage of photons, a deep fracture opened across a large portion of the proton. The fracture quickly became a gaping chasm in the topography of the proton.

The chasm had smoothly rounded edges. There were never any sharp or jagged edges anywhere. In spite of violent dynamics, there was a smoothly rounded harmony that exuded an eerie beauty. I perceived that even a single proton is an entire world, perhaps an entire universe, on some scale.

I watched as an enormous surge of electrons and anti-electrons, or positrons as they are often called, erupted out of the gaping chasm in the proton. I observed that far more matter and antimatter, that is to say more mass, came out of the proton than could account for the amount of energy that had entered into the proton in the form of individual photons of light, yet the proton did not seem to diminish in size or mass.

Later, I began pondering this process, and I wondered if this could be representative of other patterns of eternal and infinite increase. I also wondered if the infinite expanse of space itself acts as some sort of infinite reservoir from which the proton's mass and energy were replenished or reconstituted at the same rate at which the electrons and positrons erupted out of the proton following the photon impact.

A perfect system of accounting and balancing was at work in which no compulsion was required. It just worked. It was not random. It was not a fluke. The whole system was designed that way on purpose.

I understand that as electrons and positrons collide, they mutually annihilate in a burst of gamma radiation. Gamma radiation is light, energetic light at extremely high frequency and short wavelength. Perhaps the mutual annihilation that produces this light is a bit of a misnomer. Perhaps it could better be described as a mutual transformation.

Together, an electron and a positron, two mutually attracting opposites, unite and transform into a brilliant burst of light, a burst of individual photons.

Suppose an electron-positron pair originates from an eruption caused by a photon of light impacting a given proton. If that electron and that positron collide, they mutually transform into individual photons of light.

A number of those individual photons of light could then be redirected or guided back to bombard the same proton from which the electron-positron pair that had produced them originated. Then, there would be another eruption of new electrons and positrons which could then combine to produce even more photons of high-energy light, which could then repeat the process, and so on, without end, all emerging out of one given proton, all emerging out of one given world or universe, as it were.

Other protons could also be bombarded by the photons of light released by electron-positron collisions. The eruptions of electrons and positrons resulting from photon bombardment of neighboring protons could perpetuate the process exponentially and explosively, generating a never-ending and ever-accelerating cycle of expansion or increase. Seeding universes, are we?

In a way, it would seem like the electron-positron pairs originating from one world were parents with physical bodies uniting to bear innumerable spirit offspring, like a multitude of individual photons of light. Many of those individual photons of light would then go onto a world, like a proton.

While on that world, they would be instrumental in causing other sets of electron-positron pairs to arise. Some of those pairs would subsequently unite and transform to release more photons of light, and the cycle would be repeated.

At any rate, it seems that from certain vantage points, under certain conditions, it is possible to achieve perpetual energy flow, perpetual motion, perpetual increase, whatever you care to call it. It seems entirely possible, in certain systems, to get far more out than what you put in, if you apply correct principles with precise accuracy.

Perhaps an agent availing himself of these principles can achieve infinite returns because an infinite input, an atonement, which literally means at one, or unification of all things, has already been accomplished, preserving the balance of the whole system. Perhaps, if time and space and energy and matter are all permutations or arrangements of the same substance or continuum, then infinite and eternal increase can be initiated and sustained by applying certain laws of nature that are available and exclusive to the highest degree of eternal life, the degree of life that unlocks unlimited intelligence and access to the governing and the application of the entire MEST continuum.

Perhaps processes such as the photon-proton interaction that produced what I perceived to be limitless quantities of electrons and positrons are self-sustaining processes that, once set in motion by the careful application of appropriate laws, will flow forever without compulsory means. Perhaps some limited portions of those laws may be discovered and applied in lesser estates, but the whole law or the fullness of the law can only be understood and applied within the highest degree of eternal life, the only degree of eternal life where infinite and eternal increase is possible.

13 GLIMPSE OF CREATION?

On another occasion, more than a decade before the time of this writing, I had a dream in which I saw what appeared to be black, empty space, completely void of any activity and void of discernible matter. At least, that is how I perceived it.

I do not believe it was space without purpose. Nor do I believe it was space which did not belong to some intelligently governed dominion.

I saw an abrupt, intense flash of bright, white light that originated at a single point and then expanded outward in all directions as an ever expanding sphere. As the intense brightness of the initial flash receded, there appeared countless stars, densely clustered at first, but quickly moving outward from the point where the flash originated.

As the stars moved outward, they grouped into galaxies and clusters of galaxies, which in turn continued to move outward in all directions from the flash point. The stars and galaxies moved outward at great speed. A blast of gases and dust and other fine matter blew even faster, much, much faster, outward in all directions from the flash point. I saw that the gas and dust, the finer materials, reached far greater distances in much less time than the stars and galaxies.

I believe this was a representation of some phase of the creation of our universe. At first glance, it seems to lend some credence to what many call the Big Bang Theory.

However, I do not share the commonly accepted premise behind the Big Bang Theory. Instead, I believe I was viewing events from only one perspective, from a distant vantage point, looking back toward the exact location and time where the universe was deliberately and precisely placed into its element and set in motion.

I believe our universe, and every other universe, was created through various stages according to an intelligent and well ordered, deliberate design. I believe the formation of our universe, and every other universe, was placed into position and set in motion with purposeful precision and with deliberate, well organized timing, at exactly the moment intended, and not by any accident or chance or cosmic statistical anomaly.

I believe that our universe, and every other universe, was purposefully designed by intelligent, perfected human beings, many of whom are our progenitors, who have, by lawful application of their individual agency, achieved the highest degree of eternal life. I speculate that our universe, and every other universe, was conceived and organized through a certain amount of careful design and pre-assembly over a period that we, in or mortal state, would only perceive as spanning eons.

I suppose, after careful initial assembly, and after preliminary combinations of compositions and properties were arranged in the desired manner, the next step would have been what I can only describe as an ignition event, or perhaps a birth event. I speculate that some sort of ignition event, or birth event, was designed and initiated to set in motion the next phase of dynamics from which the objects and events we see today would precipitate according to the original design.

Think of it like the fireworks on the Fourth of July, Independence Day in the United States of America. You stand outside at night on the Fourth of July, anticipating the fireworks. Since it is nighttime, your eyes perceive a black expanse of sky, seemingly void of any activity or discernible matter.

Then, suddenly, you see a bright flash. Out of that initial flash, you see all sorts of beautifully colored sparks and lights and patterns expanding rapidly in every direction, some moving faster than others and reaching far greater distances than others. Then you hear a big bang (pun intended).

Does that mean what you saw was some absurdly remote accident that just happened spontaneously, by mere chance?

Did the event accidentally take place at that precise time of night, exactly on the Fourth of July, in a place that celebrates a certain event in history? Does that mean you just happened to be in the right place at the right time to witness a cosmic statistical anomaly?

Some intelligent human being had to assemble those fireworks, and then ignite them. You may have only witnessed the events transpiring after the fireworks were ignited, but that does not change the fact that some intelligence had to ensure carefully measured amounts of substances and chemicals of various properties were combined in precise arrangements within a precisely assembled vessel, including a carefully designed fuse system.

Then, some intelligent person had to ensure the whole assembly was stored in a controlled environment, preserving it against adverse elements and forces until the moment of its intended purpose. As the arrival of the anticipated moment drew near, some intelligent individual had to ensure the assembly was carefully and deliberately transported to a predetermined location.

At an exact and predetermined moment, an intelligent agent deliberately applied a precisely designed incendiary influence, like a lighter or a match, to a specific portion of the assembly, like the end of the fuse. Upon the fulfillment of these prerequisite events, the firework assembly began operating exactly as it was intelligently designed to function, having been endowed with the exact combination of characteristics necessary to allow it to fulfill its purpose, or in other words, to fulfill the intent of its creation.

Carefully composed and precisely orchestrated chemical reactions and mechanical interactions commenced explosively. While the initial explosion of the fireworks was sudden and abrupt, it followed a set of carefully planned dynamics according to an intelligent design conceived by an intelligent human being.

The fireworks explosion precipitated a beautiful display of lights according to an intelligent design. You most likely did not witness or even think about any part of this process except for the very last few events, those events occurring after the ignition of the fireworks.

You only saw the flash in the sky and the subsequent expansion of beautiful lights moving rapidly outward, forming carefully planned patterns and colors. By the same line of reasoning, is the formation of the universe really mere happenstance, a random statistical anomaly without purpose?

Can anyone honestly reason that the universe created itself without any deliberate, intelligent influence? It would certainly be more statistically probable for fireworks to assemble themselves, transport themselves, and then ignite themselves, than for an infinitely complex universe to create itself. Fireworks. Universe. Similar concepts, different scales.

Furthermore, somehow multitudes of people knew where to be and when to be there to observe the fireworks on the Fourth of July. People were generally aware of the meaning behind the event, the history it celebrates, what it represents, the freedom we enjoy, and the soldiers who fought to give us that freedom.

Might we consider that there is some significance and purpose behind the creation of the universe in which we exist? Could there perhaps be some human history worthy of celebration behind its purpose and its design, some human history that predates Earth, some human history that predates our universe? Could there have been battles fought and sacrifices made in distant places and times veiled from our minds, long before we were born, the ramifications of which we now complacently enjoy?

Immediately after I saw what I believe was the initial expansion of our universe, I saw what I can only describe as a close-up of a solar system forming within a galaxy, perhaps within our Milky Way Galaxy. I saw the formation of an Earth-like planet. Maybe it was Earth.

I saw interplanetary winds of tremendous force whipping through space, blasting the newly forming planet with dust and gases. Cumulus-like clouds surrounded most of the equatorial zone and portions of the southern regions of the planet. The planet itself appeared light bluish.

It appeared to spin with astonishing speed, as if given its momentum by the winds blowing so violently upon it. I could see no distinguishing features to suggest land and water had yet been separated on the surface. I could only see what appeared to be water and cloud bands.

I recall seeing what I can only describe as a plane of water, like a field of water in space, being blown violently and torrentially toward the spinning planet. As gas, dust, and water accumulated on the planet, I noticed that its mass and size grew.

I can only assume this was the formation of the world upon which you and I now stand. I believe it was a representation of the creation of the Earth in its pre-habitable stages, before any life had been placed upon it.

14 FROM WHENCE WE CAME

In yet another dream, again more than a decade before this writing, I saw their world. They, who? Their world was incomprehensibly vast, far exceeding the reach of any typical planetary body.

Its landscapes and beauty defied mortal imagination. I saw rolling hills as far as the eye could see, covered with tall, lush, luminescent grass, white with a subtle hue of green, undulating in a warm, pleasant, gentle, unusually hushed breeze, beneath a clear and cloudless, bright, blue sky. It was almost silent, deeply peaceful. There is no language known among men that can adequately describe such exquisite and breathtaking scenery.

Then, I saw small groups of people dressed in white. These groups interacted openly and fluently with one another, but it soon became apparent that they were distinct groups, although part of a greater whole.

It also became apparent that at a particular, though not random point in their history, each group began retiring to separate areas within the vast world they inhabited. These areas seemed to be large and unique rooms filled with features uncommon to any of the other areas. In other words, no two rooms were the same.

They were not rooms like we might envision with rectangular walls, floor, and ceiling, but rather whole distinct sceneries of various organic forms of nature, like worlds situated within the more encompassing environment of the larger, vast world.

All of the inhabitants of this vast world visited each others' rooms freely while the doors to these rooms remained open. I remember that I visited the other rooms and had friends and acquaintances in the groups that retired to those rooms, but I cannot recall any details about those rooms or the people I knew.

I only remember one room in particular. It was dimmer than the main world outside, like a deep forest with many roots and vines and enormous trees. An old wooden chair with a scalloped back sat a few feet away from an equally aged wooden picnic table beneath a large tree that looked like a giant sequoia.

As a tall, thick, wooden door covered with vines was slowly being shut, I remember there was some conversation with inhabitants of the neighboring areas, as if farewells were soon to be exchanged. I cannot recall the exact conversations.

The only words I recall, I do not fully understand. These words were spoken within the group of which I was a member. They were uttered as the door was shut.

Someone addressed one of the girls of this room and said something to the effect of: "... so take us beyond these walls." Perhaps the girl was Eve, the mother of the human family that pertains to our Earth.

When the door was shut, it sealed, seamlessly integrating with the surroundings in such a way that it was no longer discernible as a door. The inhabitants of the room would be unable to locate it or recognize it, let alone open it. It seemed as though it was only opened and shut from time to time by someone from the vast outside world who had the ability to operate all the doors of all the areas or rooms to which the various groups of inhabitants retired after communing in that spacious world outside.

Before the door to the room to which I pertained was shut, I had noticed that eventually, the opening of the doors to the various rooms, and the interactions among the various groups, became less frequent and shorter in duration. Each group seemed to spend less and less time outside in the vast world communing with fellow groups.

Though nothing was said explicitly, it felt as if each group became increasingly preoccupied by some significant event that was going to take place, an event that would have an everlasting effect upon every individual member of their group, as well as upon their group as a whole. There was an unspoken realization that very soon nothing would ever be the same.

Each group seemed to be preparing for a similar type of event. But, for some reason, there seemed to be such uniqueness and such urgency about these impending events, that eventually the groups became so occupied with their own specific preparations that they had no time to discuss anything with neighboring groups. Not that there was any lack of interest or lack of concern for what other groups were doing or about to experience, but there was simply no time to explore and share the details with other groups.

Finally, the time came when the doors of each room were shut and sealed. Subsequently, the groups neither saw nor heard from anyone in any of the other groups, as far as I could tell. At least, I recall that being the case for the group to which I pertained. In this group, the very memory of all the other groups began to fade, so engrossed were we in our own urgency.

I assume, perhaps erroneously, that the same type of situation overcame the other groups as well, and that in all likelihood they would not be able to remember us, nor anyone from any group aside from their own. The thought crossed my mind that unbeknown to us, many if not all the other groups could at some point be restored to some degree of interaction with each other, but not with my group until some distant time.

It also crossed my mind that my group was, for some reason, specifically prohibited from retaining any awareness or interaction with the other groups, at least until some distant time, and after a series of specific events to which my group alone must be subjected. But these thoughts are mere speculation on my part.

Perhaps time is nonlinear. Perhaps time has geometric dimensions just as space does. Perhaps there are cases where history not only repeats itself but parallels itself in separate and distinct timelines.

Time flows at different rates and is perceived differently within different frames or spheres. Perhaps there are nonparallel timeframes, some intersecting and some nonintersecting.

Perhaps all of the groups of people on the vast world shared some generalities in the types of events they would face. The inhabitants may have been temporarily veiled from any awareness of details outside their own course of history, outside their own frame of reference, until they would eventually come to some point in their history where a restoration of all awareness would become appropriate.

The inhabitants may have perceived and measured time in completely different ways, depending on the worlds to which they were assigned for their mortal probations. What we may perceive as unimaginably prolonged eons of waiting for some predicted event may amount to a few decades or a few centuries or a few millennia when considered from the perspectives of peoples assigned to different worlds.

For example, say you had a close friend who shared much in common with you. Suppose you each had to move away, and you lost contact with each other for many decades. It is plausible that each of you could have been experiencing similar types of events in your own lives, unbeknown to each other.

The specifics, the timing, and the chronology may have differed, but the general kinds of events were likely similar. Perhaps each of you graduated from some institute of learning, engaged in some form of livelihood or occupation in your chosen fields, pursued hobbies and interests, raised your own families, endured tragedies and hardships of health and economics and failed personal relationships, and relished the joys of successes and triumphs and harmonious relationships.

All the while, neither of you was even remotely aware of what the other was doing, but a distant observer who could see the course of both your lives could clearly see the parallels in your experiences, and in the experiences of countless others. Then, after some period of time, which may be drastically different for each of you, both of you reach a point where you are able to reunite and relate your respective experiences to each other, and describe to each other or show each other the respective worlds in which you had lived.

One of you may feel you had not seen each other for decades. The other may feel as if you had only been apart for a brief moment.

Consider snowflakes. No two snowflakes are exactly the same, but they share many common traits. They are all made of water that has been temporarily frozen as ice under similar sets of conditions. They are all clearly identifiable as snowflakes. They are all subject to wind and gravity and temperature, regardless of where or when or in what order they precipitate, or how fast they precipitate.

Likewise, no two people are exactly the same, but they share many common traits that make them clearly identifiable as human beings. They are also subject to similar types of influences and conditions, regardless of what world they inhabit, or when, or in what order, or for how long they live in a mortal tabernacle.

I recall that the room in which my group lived was the dimmest and most enclosed of all the other rooms I had visited, though I have no specific memories of other rooms. I can remember as much about the vast world outside the rooms as I can about my group's room, yet I have no recollection of the details of any of the other rooms.

I feel that at some point in history, various groups, including the one to which I pertain, will rediscover each others' existence. We will eventually restore mutual interactions and recognize that we all came from the same place, and that we have common progenitors.

As I pondered all this, it soon became obvious to me what was being represented. The vast world was the place where we lived as spirits before we were born into our mortal bodies on this earth. The vast world was the place where we lived as spirits before our earth was created. The rooms represented various different planets in their formative stages of planning, design, and creation, prior to their respective inhabitants being born onto those worlds to begin their mortal probations.

I understood that the groups of people dressed in white were the human inhabitants pertaining to those planets before they were born onto their respective planets. It appeared that we could freely visit each others' worlds during their various stages of design and creation, and that we were all familiar with each other as members of a great, extended family, originating from parents who lived on the vast world in which the rooms were situated.

Apparently, we all knew each other quite intimately, regardless of the fact that we were organized into groups pertaining to unique and separate worlds. Our communion with each other was frequent and open at first, but as the time drew near to finalize the creation of the various worlds, and then begin inhabiting them by being born onto them, each group necessarily became so occupied in their particular activities that the doors to the rooms were less frequently opened, and our visits outside in the vast, common world, and to each others' worlds, became less and less frequent and of shorter and shorter duration, until finally the doors to the various worlds were shut and sealed.

It seems that the doors were not all shut simultaneously, but they did seem to be sealed within the same general era, according to some previously planned timing.

Not long after the doors were sealed, I recall people beginning to depart from the room my group inhabited. Their departure was extremely swift.

Each individual that departed was physically whisked away in a blur, disappearing from view altogether. This happened within a relatively short distance, as if you were standing a few yards away from someone, and then suddenly that person moved as fast as your eye could track while simultaneously fading in a translucent blur, diminishing in scale while accelerating out of view, toward a vanishing point.

I believe such departures from the room represented births into mortal life. Births as viewed from the perspective of those yet to be born.

Up until their birth, these individuals were spirits, in their prime, waiting to be born into mortal bodies to grow, live, and die on the world that had been created for them.

Upon the death of their mortal bodies, those same spirits would inhabit a temporary holding area, like a waiting room of sorts, a spirit world pertaining to, and existing upon, the same world into which they were born as mortals. I do not recall seeing that spirit world.

I do recall one detail common to all individuals before they departed to be born into their mortal bodies. They did not reflect light. Nor did they cast or receive shadows. They were luminous in and of themselves. They radiated their own light, and they were discernibly brighter than the ambient light.

I did not notice at the time, but in my own speculation later, I imagined if I would have more closely examined the events before me, I should have noticed that the intensity of light radiated by various individuals was not equal. All individuals emanated their own light, and that light shone with an intensity I would have assumed was somehow related to their several degrees of individual development up to the point at which they were to be born into mortal life.

15 TOROIDS AND CHAIN LINKS

I don't know how or if this fits into the puzzle, yet, but I have been pondering the idea of toroidally shaped concentrations of energy, somewhat akin to the rings mentioned in Superstring Theory. I do not believe such vibrating rings of energy are truly elemental or fundamental.

Perhaps, if such rings could be examined more closely, we might notice they are composed of ever smaller toroids. I think the rings of energy might be composed of toroids that have at least the following types of dynamics:

Picture a donut whose surface flows about a circumferential axis. Let's say we view the donut from above. The surface flows inward from the outer circumference to the inner circumference. The sprinkles on the frosting are moving from the outer edge toward the hole in the middle, then down the hole, then around the bottom and back up the outer edge, over and over again.

Now suppose the surface is flowing in the above described manner at a different angular velocity than the material below the surface. There is now a shearing action between any given depth and any other depth of the material. That is to say, if you cut the donut such that you have a circular, cross sectional slice, and you look along the axis perpendicular to that circular slice and going through the center of the slice, you might observe swirling currents with differing angular velocities at different radial distances from the center of the circular slice.

Now, suppose not only do the swirling currents vary with distance from the center of the circular slice, but they vary from one slice to the next. So, there are shearing actions between any two circular slices.

Suppose the swirl at any given point is not steady, but changes angular velocity, and even reverses direction. In other words, there are angular oscillations.

Zoom out and look at the entire donut. Imagine, in addition to the internal dynamics just described, the whole donut is spinning about the axis running through the center of its hole.

As it spins, suppose the cross sectional areas of the circular slices are changing, such that the donut gets thicker and thinner in wave patterns propagating in the direction of spin (prograde). Suppose there are retrograde wave patterns superimposed on the prograde thickness waves.

While all that is going on, imagine undulations displacing the circular cross sections of the donut in directions both perpendicular to and parallel to the axis about which the entire donut spins. Mutually perpendicular waves. Sounds a bit like electromagnetism. Electric fields are perpendicular to their associated magnetic fields.

Of course, the cross sectional slices are not circular at all, but elliptical. That adds yet another set of variables to the dynamics, as elliptical cross sections can vary in both size and shape.

As suggested in Superstring Theory, the frequencies of vibrations in these toroids may define material properties. Superstring Theory suggests space is composed of arrays of discrete, vibrating rings.

But, what if these donuts, or toroids, are not discretely separated? Suppose one donut is linked to another, like links of a chain. They can be linked in a line, like a single chain strand. They can be linked to form a sheet, like chain mail armor. They can be linked to form a cube or any three-dimensional space, like a matrix or a sponge. This can go on and on. They can be linked in n-dimensional space.

Now, suppose some of these interlinked toroids in n-space are tightened or shrunk in such a way that they organize space itself into a material particle, a mass. The deformation of adjacent, linked toroids is the warping of spacetime we would expect around concentrations of mass (or energy).

Falloff for attributes such as luminous intensity and gravity varies with the square of the distance from the center of the concentration of mass-energy, because that is the way whoever designed and set up the dynamics of the toroidal constituents wanted it. Because the values mesh to enable what is intended to be, to be. And the values mesh to prevent whatever is not intended to be, from becoming.

There are, of course, any number of possible combinations for linking an infinite number toroids together. There may be many toroids all linked to any given toroid, like keys on a key ring.

There may be toroids linked in simple chains. There may be chains that curve to form larger toroids, which are in turn linked as chains, and so on. There may be spirals and helical arrangements analogous to DNA. Remember, however, we are talking about scales smaller than the size of a quark.

Perhaps, as suggested by Superstring Theory, such structural patterns, together with their endowed dynamics, define the properties of everything in the MEST continuum. Not only might the various patterns and their dynamics define things, but they might also be the means by which things are bound within the dominions to which they are consigned, or to which they have consigned themselves, when speaking of that which acts intelligently, having agency.

Could the very fibers of our spirits be arrangements of such patterns and dynamics? Could this be the means by which intelligence is integrated with spirit matter, rather than merely housed inside it like a hand inside a glove?

Could interlinked toroidal MEST formations be the means by which our spirits and our physical bodies interface? The nature of spirit-to-body interface is temporal and terminable in the case of mortal bodies, but wholly and permanently integrated and inseparable in resurrected bodies. Resurrected persons have the ability to reveal themselves to, and conceal themselves from, mortals. Are there fibers of light and fibers of darkness?

16 HOW BIG IS A PHOTON OF LIGHT?

What is the size of a single photon of light? Perhaps that is a relative question. How big is a single photon, relative to what? Perhaps the answer depends on how small a scale we want to consider. In other words, how small do we want it to be?

Given the dual nature of light, we can either speak of the size of a photon as a particle of light, or as the wavelength and amplitude of a wave of light. Either way, we can zoom in on a single photon to infinitely miniscule scales and still have light. If we consider a photon as a particle, we would likely find it composed of smaller particles, perhaps shaped somewhat like vibrating toroids, each of which would likely be composed of even smaller vibrating toroids, and so forth.

If we consider a photon as a wave, we can imagine waves on the wave, somewhat like water has waves on waves on waves, ad infinitum. But, there is another way to look at the photon as a wave.

When we ask how much energy a single photon has, we can get an idea of the size of the photon. If the speed of light in a given medium is constant, then speed has nothing to do with variations in the energy of a photon.

If one photon has less energy than another, the one with less energy must have some combination of lower amplitude and lower frequency than the other. But, wait. Max Planck discovered that electromagnetic energy, light, in our universe anyway, behaves as if it were discrete packets of energy.

Consider the relationship, $E=hc/l$, where E is energy, h is plank's constant, c is the speed of light, and l is wavelength. Note that amplitude is conspicuously absent from this equation. Why? I am guessing it is because Planck's constant shows that there must be a constant amplitude for light. That being the case, the only factor that can vary infinitely is wavelength.

The less energy a photon has, the longer its wavelength. This means the photon with less energy is actually bigger, or longer, than the photon with greater energy. The greater the energy, the shorter the wavelength.

That is why mass shortens along its velocity vector as it dilates approaching the speed of light. That is also why mass dilates and flattens along an axis normal to the spatiotemporal curvature near a singularity, relative to fixed observers afar.

This suggests that, relative to a fixed observer, nothing can exist in less than three dimensions and have finite mass or be subject to time. Does Planck's constant suggest that mass, like light, has a dual nature as a wave and a particle? Energy is mass and mass is energy, and light is energy, so mass (matter) must have at least a dual nature as a wave and a particle. Since mass curves spacetime, curved spacetime represents mass. So, in a sense, mass is a concentration of space and space is a concentration of mass. The MEST continuum.

Planck's constant is **h=El/c**, so **l=hc/E**. Since **E=mc²**, we can express wavelength as **l=h/(mc)** and mass as **m=h/(lc)**. This further indicates that matter is light and has at least a dual nature as a particle and a wave. Mass must dilate as its wavelength shortens, relative to fixed observers. Wavelength must decrease as mass increases, relative to fixed observers.

Can we look at this as mass expressed as a one-dimensional component of space? Any given wavelength of light, which is a measure of some one-dimensional component of space, can be expressed as mass. This goes along with the concept of the MEST continuum.

Another interesting fact discovered by Max Planck is that action must always occur in discrete packets, or quanta. While many things can be infinitely and arbitrarily adjusted, action cannot. There are minimum packets of action small enough such that there can be no smaller action, no partial action.

In other words, while we can dissect mass, energy, space, and time into smaller and smaller pieces, to infinitely small scales, there is a definitive scale beyond which we cannot subdivide action into smaller constituent actions. This implies that, ultimately, partial actions are forbidden or impossible.

This has eternal ramifications in the material and the spiritual sense. Choices can be made by intelligence, because intelligence has agency to act. However, specific consequences cannot be separated from specific exercise of agency.

No intelligent agent can commit less than a quantum of action and expect a quantum of results from that action. No less than a quantum of consequential action can result from a quantum of chosen action. That is a law irrevocably decreed by the intelligence that organized for us our universe.

This is why, in the eternal perspective, there are discrete degrees of glory, or kingdoms, as it were, and not just one heaven and one hell. This is why it is said that, spiritually speaking, a person must be hot or cold, not lukewarm. This is why there is a Gulf of Lazarus that prevents inhabitants of lesser dominions from accessing higher dominions. This is why where some are, others cannot go.

This is why the ordinances that have been established to enable us to achieve whatever degree of eternal life we desire must be completed in their entirety. An incomplete ordinance is the same as no ordinance. Partaking of the bread or the water avails nothing. One must partake of both the bread and the water.

A partial tithing secures no promise whatsoever, while a full tithe literally opens the windows of heaven. This is also why sacred covenants afford constant protection while the abandonment of such exposes an agent to eternally destructive influences.

Why, in a universe of infinitely many variables, are there a mere two dozen or so known constants? Of course, there must be constants we have yet to discover. But, why so few constants amid infinite variability?

This applies again, to both the material and the spiritual. There are but few constants. However, those few simple, little constants anchor all things in there respective spheres, and enable the existence of all things.

Those few constants uphold the framework within which intelligence can act and have access to that which is acted upon. Those few constants ensure that that which is purposed to be can become and that which is not expedient does not become.

Constants enable the orderly delineation of that which acts intelligently, according to the laws that intelligence chooses to abide. Constants ensure that that which is acted upon by intelligence is allocated commensurate with the agency exercised by the intelligence. There is order amid an infinite sea of variables because of a few simple constants.

Now, here is something that makes me want to jump out of my chair and shout, "Gnarly!" But I might hurt myself, so I won't.

Given the fact that lower energy photons are longer than higher energy photons, we can boldly connect that which the secular world has never before connected.

Intelligence is light. That may not be common knowledge in the current secular world, but it is nonetheless a fact that has been authoritatively established. At some point it must be considered in the quest for a unifying equation.

The fact that lower energy photons are longer than higher energy photons is consistent with the idea that lesser intelligences are incapable of resolving and influencing highly refined details in the MEST continuum. The finer the detail an intelligence desires to observe or influence, the smaller the wavelength of light it must apply.

Small and simple means effect great things. The shorter the wavelength, the higher the frequency and the higher the energy.

This, in turn, is consistent with the idea that an agent within the frame of a luminous singularity, that is an infinite concentration of light, in other words an agent possessing infinite intelligence, would have all things, great and small, past, present, and future, before him simultaneously in one complete whole.

There would be nothing too small to be observed or influenced by such an agent, because he can apply infinitely short wavelengths of electromagnetic energy, which is another way of saying infinitely high frequencies of electromagnetic energy, which is another way of saying infinitely high energy. In other words, he has infinitely powerful thoughts applied with infinite resolution, such that nothing escapes his awareness or his influence.

And since such an agent has infinite energy, and thus infinite matter, effectively the whole MEST continuum at his disposal, he can afford to transport any amount of it anywhere in the cosmos, and he can do so instantaneously. Is this how stars get some of the energy they radiate, or some of the antimatter they annihilate to supplement their on-site fusion?

Is this how miracles happen in our personal lives? Is this how we can be in, and interact in the same dreams with each other? Is this how astral projection, telekinesis, and telepathy work? Is this how prayers and answers to prayers are transmitted?

It seems this layman's small and simple path of reasoning has come full circle, circumscribing a hypothesis that may begin to explain how everything works and how everything is interconnected. I did not expect that the line of reasoning with which I started out in the first chapter of this work would bring me full circle to this chapter about the size of a single photon of light and how that relates to intelligence.

Maybe this means someone in a frame of a luminous singularity is transmitting hints to my relatively uneducated mind, trying to educate me, line upon line, precept upon precept, until, eventually, I will understand how the whole puzzle fits together.

17 BOLD LEAP FORWARD

In brief summary, the Circumscription Hypothesis replaces the Uncertainty Principle with a Certainty Principle and suggests a quadruple nature of all things. Everything exists on the MEST continuum. The MEST continuum is the whole within which everything can be explained.

There is one unifying equation that accounts for everything. We have not yet discovered it in the mathematical language. However, I believe the principle of it is literally embodied in the Atonement of Jesus Christ.

I believe such an equation, once expressed in the mathematical language, will be derived from an understanding of the MEST continuum and the quadruple nature of all that exists. I further believe that a unifying equation must acknowledge and factor in the two overarching castes of the MEST continuum: things that act intelligently and things that are acted upon by intelligence.

The Circumscription Hypothesis and the Electric Universe Theory are mutually supportive. The Circumscription Hypothesis is a bold proposal. I believe it will, at some future date, become a viable and recognized theory. Eventually, it will become an irrefutable scientific fact. It will prove empirically what can only be learned by faith at present.

ABOUT THE AUTHOR

Ilyan Kei Lavanway is a Christian, a family man, and a deep thinker. He has had many dreams that he believes have significance in his quest to comprehend the workings of God in his personal life and upon the cosmos as a whole.

Mr. Lavanway is fascinated by the Creation and Christian eschatology. He seeks to understand and worthily participate in the fulfillment of prophecies and events of the upcoming days preceding and following the Second Coming of Jesus Christ.

Mr. Lavanway served as a full time missionary for two years in the Argentina Buenos Aires North Mission of the Church of Jesus Christ of Latter-day Saints, 1986 to 1988.

He feels he can continue doing missionary work through his writing.

Mr. Lavanway is a United States citizen and a patriot. He loves his country and the U.S. Constitution as envisioned by the Founding Fathers. Mr. Lavanway served on active duty in the U.S. Air Force for almost fourteen years in various non-flying operational and academic assignments.

While he never flew for the Air Force, he has always harbored a passionate interest in aviation. Since his circumstances over the past two decades have made it impractical for him to fly, Mr. Lavanway has taken up writing and discovered that he loves it almost as much as he loves flying airplanes.

It is his hope and prayer that some of his written works may serve to open minds and inspire readers. If any portion of his works influence others in their own personal quests for truth, it has been effort well spent.

Other books by Ilyan Kei Lavanway

An Aviator At Heart (2014)

Intelligent Universe (2014)

Sevenfold (2013)

Post Omerican Easter (2012)

The Modern Day Gadianton Golden Boy (2012)

Out of the Picture and Into the Picture (2012)

Platypus Boy on the Duck Farm (2012)

Duck Boy on the Platypus Farm (2012)

Earth Sink (2010)

Connect with the author

Email:
ilyanlavanway@yahoo.com

Amazon author page:
http://www.amazon.com/author/ilyan

Goodreads author page:
http://www.goodreads.com/ilyan

Smashwords author page:
https://www.smashwords.com/profile/view/ilyan

Self Publishers Showcase page:
http://selfpublishersshowcase.com/ilyan-kei-lavanway/

Book Crossing page:
http://www.bookcrossing.com/mybookshelf/ilyan/all

Facebook:
https://www.facebook.com/ConspiracyParanormal

LinkedIn:
http://www.linkedin.com/in/ilyanlavanway

Twitter:
https://twitter.com/ilyanlavanway

Blogs:
http://ebooksscifi.wordpress.com
http://conspiracyparanormal.blogspot.com

Website:
http://ilyanlavanway.wix.com/books

www.ingramcontent.com/pod-product-compliance
Lightning Source LLC
Chambersburg PA
CBHW030916180526
45163CB00004B/1850